GOAT TAGS

The Rise of Cyborgs and
the Mark of the Beast

Peter John Brooks

GOAT TAGS: The Rise of Cyborgs and the Mark of the Beast

Scripture taken from the New King James Version® unless otherwise noted. Copyright © 1982 by Thomas Nelson. Used by permission. All rights reserved.

Published by Fivestone New Media
Print ISBN: 978-1-968804-13-8
Digital ISBN: 978-1-968804-12-1

www.bethelcornerstone.org

Computers are going to keep getting smaller and smaller. Ultimately, they will go inside our bodies and brains and make us healthier, make us smarter.

Ray Kurzweil, Google (*Kurzweil*, Time)

He causes all, both small and great, rich and poor, free and slave to receive a mark on their right hand or on their foreheads, and that no one may buy or sell except one who has the mark.

Revelation 13:16-17a

Contents

Introduction

See, I have set before you today life and good, death and
evil.

Deuteronomy 30:15

HUMANITY IS REACHING a critical juncture.
Moments like this don't come along often, but
when they do, they mark grand turning points in
the human story. In the past two centuries, technology has sparked great change. The industrial
revolution, steam engines, automobiles, airplanes,
and now computers and the internet have all in-

credibly altered the human experience. But now we are at the brink of a transformation that may be greater than any other.

This revolution might be called the *internalization of technology*. The technology that we rely on for daily tasks— inside our phones, laptops, GPS navigators, and other devices— is about to take the big leap from being inside external objects held in our hands to being inside our own bodies.

Information systems have already become pervasive, entwined with almost every aspect of life. We plug in for communications, shopping, information, socializing, and entertainment. For many people today, living without a smartphone or computer is almost unimaginable. Information technology has invaded our homes and cars, making them "smart," enabling them to communicate with each other and with powerful external computers. This "smartness" is taking over our refrigerators, lights, and thermostats.

Our experience of technology is about to get much more intimate. Soon this gray advance of computer algorithms will be attempting to invade our bodies, posing, of course, as a helpful and even necessary development which will aid us in our pursuit of health and happiness.

Technology has already found entrance into *homo sapiens*. Microchips are not just for pets anymore, to identify them if they run off. Some people are chipped in their hands— to unlock a door, they place their hand near a sensor. To pay for a snack, they wave their hand near the vending machine, and the money is debited from their account. Brain-machine interfaces assist stroke victims to speak and move.

Currently, these internal uses for technology are isolated and specific, and a pacemaker does not a cyborg make. Sea change will happen when internal technology is ready to aid us with more mundane matters like identification, banking, communication, mapping, or information access. As internal technology grows, it will replace computers, phones, and even doctors. People will plug in directly to powerful computers and be transformed.

GOAT TAGS

We are accustomed to technological upgrades being beamed into our devices. But in a few short years, the next big upgrade to our favorite devices will be the internalization of these devices into ourselves. When this happens, humans will become cyborgs.

This future will be radically different. These emerging technologies will have profound implications for all of humanity, especially for Christians. It's time to understand and prepare.

Chapter 1
Here Come the Cyborgs

A prudent man foresees evil and hides himself, but the simple pass on and are punished.

Proverbs 22:3

CYBORGS ARE COMING. The most powerful organizations on the planet are working hard to bring cyborgs into existence. Recent developments in technology indicate they are just around the corner. Experts claim that cyborgs will be the next stage of human development - homo sapiens 2.0.

According to the Merriam-Webster Dictionary, a cyborg is "a person whose physiological functioning is aided by or dependent upon a mechanical or electronic device."

In October 2017, the Parliament of the United Kingdom heard testimony from experts who stated: "Within just two decades, technology may have advanced so much that humans and machines are effectively 'melded' together" (*Knapton*).

This melding between humans and machines might look like this.

Physically, cyborgs might have bionic limbs, giving them superhuman strength. They will be able to run faster, hit harder, and endure extreme environments. Humans get tired and have strength limits, but if their bodies are combined with machines to make them stronger and faster, they will be able to break through these limitations.

Cyborg eyes will be enhanced through electronic overlays, enabling them to see computer-generated images superimposed onto reality around them, providing immediate information about their surroundings. Digital pop-ups that will appear on these overlays will provide details about specific persons whose faces they see on the street.

Cyborgs will have health advantages over the merely biological. Nanotechnology might swirl around in their bodies, identifying and solving health problems before they even start. Implanted technology will potentially give cyborgs ease of managing diabetes and solving dreaded diseases like heart disease and cancer.

Mental enhancements will enable cyborgs to have a direct connection between their brains and the internet, giving them the ability to access online information at the speed of thought. With minds connected directly to computers, cyborgs will be much more capable of solving problems and facing complex situations than normal humans. They might be able to communicate with others almost telepathically, making phone calls using their thoughts and carrying on conversations without external devices. They will have immense advantages in education, research, or in the workforce.

Cyborgs are poised to change the world.

This is not science fiction. Technology is already being implanted into the human body. Pacemakers work with the electrical currents of the heart, enabling it to beat normally. Cochlear implants assist in hearing, not by actually producing sound, but by digitally stimulating a nerve with a small computer. Implants regularize the brain activity of victims of Parkinson's disease. Thousands of people around the world have already been implanted with microchips which offer identification or payment capabilities.

Implants like these do not necessarily turn people into cyborgs, but when enough technology gets inside the human body so that what it means to be human is redefined, then cyborgs will arise.

Growth of Technology

Technological growth over the past 150 years has been astounding. Take something as common as the telephone. Decades ago, immediate communication by voice across long distances was seen as impossible. Predictions about the impending arrival of the telephone were met with blank stares or scoffing. But while other people were going about their normal lives, Alexander Graham Bell was busy in his laboratory, experimenting with electrical impulses in wires. In 1876, his groundbreaking technology was ready. He hooked up his wires, used them to speak to someone miles away in another city, and the telephone was born. This invention changed the world, and soon, telephones became part of normal life.

Today, there are many well-funded laboratories similar to Bell's early phone lab that are exploring the boundaries of internal technology, nanotechnology, robotics, and artificial intelligence. Some of the progress made in these feverish laboratories has given astute observers the chills. Experts as diverse as physicist Stephen Hawking, businessman Elon Musk, and a host of professors from institutions such as Cambridge, Oxford, Stanford, Harvard, and MIT are sounding the alarm bells about the direction this technology is heading (*Research Priorities*). This research is all conspiring to turn you and me into humanoid machines. One day, something will burst out of one of these laboratories that will completely change our experience of the world.

Let's go back to the telephone. This technology didn't stop with Alexander Graham Bell. In 1973, Motorola made history when the first phone call on a cellular phone was made. Suddenly, people gained the freedom to carry their phones and their conversations almost anywhere, without being tethered to wires. Today, this once "impossible" technology is taken for granted, and now 86% of people in the world have access to a cell phone, more than have access to a toilet.

Phones have gone from being huge contraptions only at public buildings, to being on the walls of our homes, to being in our pockets, to being on our wrists. Where will they go next? It doesn't take a genius to predict they are going inside our hands or heads.

Limited expectations regarding technology and how it will affect our world are often exceeded as new discoveries and developments are made. Things like telephones, computers, airplanes, cars, the internet, and spaceships all once seemed like incredible fantasies to most people, but soon after their invention, they became accepted parts of culture and society. Humanity quickly forgot how unimaginable they once seemed.

Technology is changing our lives at a pace probably never seen before, yet we are still limited in understanding how technological change will affect us in the future. There are many technological possibilities now that seem impossible to the average person. These new technologies, when mentioned prior to their arrival on the mass market, are often met with skepticism.

Most people find it very hard to believe that their neighbors might become cyborgs. It's like television in the 1920s. Most people just couldn't believe that a box would sit in their houses and show pictures and sounds of things happening thousands of miles away. It seemed unimaginable. But as soon as TVs came into homes and families began to sit around watching the evening news, the impossibility of it all was quickly forgotten. Decades later, it now seems as if TV has just always been there.

It is similar with other new technologies: cars, electricity, lights, radios, microwave ovens, GPS navigation. These technologies seemed to flash out from nowhere and quickly permeated society. Suddenly they appeared on the store shelves, came into houses, and became part of normal life. Not only are these technologies now taken for granted, but they are seen as so essential that life without them is suddenly unimaginable.

Exponential Revolution

There is much more to come. Technological advances are increasing exponentially. Computers drive modern technological change, and they keep getting faster, smarter, and more powerful. For the past 50 years, the processing speed of computers has been doubling approximately every 18 months.

This rate of increase is called Moore's law, named after Gordon Moore, one of the founders of Intel, who first predicted this rapid pace of technological advance in 1965. Over 50 years later, Moore's law still holds true, and it will probably hold true for many years to come. This law is a reliable indicator of general technological increase, as improvements in computer speed spur advancing technologies which in turn transform our culture at a rate probably never seen before.

Coping with technological change has become a normal part of our lives. In order to keep up to date (maybe even to stay sane) in this time of rapid change, we have learned to quickly adapt. Our gadgets are continually being upgraded, becoming lighter, slimmer, faster, more powerful, and more intuitive. We change our habits along with them. When using passwords is the norm, we use passwords. When scanning our fingertip is the norm, we scan. When typing on a keyboard is the norm, we type. When speaking to Siri is the norm, we speak. When powerful companies urge us to migrate our storage from local hard drives to remote servers on "the cloud," we do just that, not really considering the implications.

We have learned to adjust to new technology as if to changing fashions. When new computers or phones come out, we want to buy them. When an update to our operating system arrives, we don't think twice before upgrading. We don't have time to read the fine print. We click on the box on the screen and accept the terms and conditions without even really knowing what they are. We accept the promise of new technologies to make our lives easier, smarter, faster, and more fun.

Constant improvements are making the interaction between user and device more and more seamless. Ease of operation is the goal. With each improvement, the gap between the user and the device is becoming smaller and smaller. Our phones have almost become an extension of ourselves. We're so attached to them that the average American now spends over 10 hours every day looking at a screen (*Sheikh*).

We have become very close to our beloved devices. In spite of this intimacy, a gap still remains. We must speak to them, poke at them with our fingers, stare at them, pick them up, carry them, put them in our pockets or purses, and do other things that emphasize that, as intimate as we are with them, they are still external to our persons. This gap will remain as long as technology sits inside devices external to the human body.

This gap won't remain for much longer, because this electronic stuff is going inside us. This will be good news for hardcore technophiles. When technology is part of our bodies, we won't have to stare at our screens all day and tire our fingers through typing. Technology will become a part of us, accessible through our thoughts and in front of our eyes at all times.

Humans and the technological devices they love are about to be united together, married in a mechanical, biological union. The fruit of this union will be the cyborg.

Closing In

Powerful corporations are working on transforming humans into cyborgs. One of the most active is Google, one of the largest and most influential corporations in the world. Their popular operating system for smartphones is named Android, another name for a humanoid robot. Android now runs nearly 90% of new smartphones. Google's efforts were ideologically spearheaded by Ray Kurzweil, who became Google's Director of Engineering in 2012. He dreams of uniting people with machines so they can live forever.

Kurzweil shockingly said, "We're going to become increasingly non-biological to the point where the non-biological [machine] part dominates, and the biological part is not important anymore. So even if that biological part went away, it wouldn't make any difference" (*Kurzweil,* Fantastic Voyage).

"Even if that biological part went away"...

Kurzweil is basically suggesting that we become robots, completely lose our humanity in the process, and that such a development won't be bad at all.

In other words, eliminate humanity through the implantation of technology.

Cyborgs then, for Kurzweil, will be a temporary stop along humanity's journey toward becoming androids - pure robots.

These are not the wide-eyed dreams of a science fiction writer. These are the measured goals of one of the wealthiest and most influential corporations on earth. The goals of other powerful technology companies are not terribly dissimilar.

It's not just wealthy corporations that are searching for cyborgs.

A Hardware Update for the Human Brain: From Silicon Valley startups to the U.S. Department of Defense, scientists and engineers are hard at work on a brain-computer interface that could turn us into programmable, debuggable machines. (*Mims*)

The Government of the United States, probably the largest and most powerful organization in the world (besides the true church), has also been working on this for years. What is developed in the halls of US advanced laboratories often trickles down into the daily lives of ordinary people. That's what happened with the internet, originally developed as a project with the US Department of Defense. The internet burst out of nowhere about 30 years ago and is now a regular part of most people's lives. The US Defense Advanced Research Projects Agency (DARPA) has been working on ways to implant computer chips into human brains for many years. There are some reports that they have already begun implanting chips into the brains of soldiers.

DARPA has implanted a chip in the brain of a subject who, using only his brain, has controlled three drones simultaneously.

"Using a bidirectional neural interface, a volunteer named Nathan Copeland was able to simultaneously steer a simulated lead aircraft and maintain formation of two simulated unmanned support aircraft," said Tim Kilbride, a DARPA spokesperson.

Not only did Copeland send signals to the drone, the drone sent signals back.

"The signals from those aircraft can be delivered directly back to the brain so the brain of that user can also perceive the environment," said Justin Sanchez, the director of DARPA's Biological Technologies Office. (*David*)

This experiment involved a "bidirectional" interface. Not only did human thoughts control the drones, but information from the drones was transmitted directly into the subject's brain.

Let's get our heads out of the sand and face the future squarely with clear eyes. The transition of humans into cyborgs is something that is going to happen right in front of our noses (possibly *to* our noses or other organs), gradually, piece by piece, step by step. First there will be bionic limbs for amputees. Next there will be internal medical devices to treat diabetes or heart disease. Nanotechnology will swirl around in the blood to detect and prevent disease. Eyeglasses will be replaced with computerized glasses which will superimpose computer images onto our view of the world around us. Contacts will be replaced with an electronic overlay to the human eye. Identification markers inside our hands will tag us for logging into the internet or making payments. Our brains will be implanted with chips that will enable us to access powerful computers, plugging us directly into the internet.

Step by step, implanted technology will begin to take over humanity.

"Professing themselves to be wise they became fools, and changed the glory of the uncorruptible God into an image" (Romans 1:22-23a).

Chapter 2
Plugging in the Human Mind

Thou canst not touch the freedom of my mind.

John Milton, *Comus*

COMPUTERS ARE LEARNING to read the human mind. Brains emit electrical signals as neurons fire, and computers are beginning to understand these signals. Brain-computer interfaces convert brain information into electronic data, turning thoughts into bits and bytes. This will ultimately allow humans and computers to communicate using a common language, allowing information to smoothly transfer from the human brain to computers and vice versa, along a two-way information highway. Remember the "bidirectional interface" from DARPA? Such interfaces are being developed in other labs too.

> "We're popularizing the use of BCI (brain-computer interface) instead of it being stuck in the research lab," said Chris Crawford, a PhD student in human-centered computing. "BCI was a technology that was geared specifically for medical purposes, and in order to expand this to the general public, we actually have to embrace these consumer brand devices and push them to the limit."

Scientists have been able to detect brainwaves for more than a century, and mind-controlled technology already is helping paralyzed people move limbs or robotic prosthetics. But now the technology is becoming widely accessible.

Unlocking a car or exploring a virtual world could one day be tasks achieved by thought alone. It could also be applied for real-time monitoring of our moods and states of consciousness. One day you could wear a brain-controlled interface device like you wear a watch, to interact with things around you.

"The progress of the BCI field has been faster than I had thought ten years ago," said Dr. Bin He, biomedical engineer at Minnesota University.

"We're getting closer and closer to broad application" (*Dearen*).

Powerful companies are working to make brain-computer interfaces a widespread reality, with broad consumer application.

The U.S. Department of Defense is seeking to "develop a 'neural interface' that would both allow troops to connect to military systems using their brainwaves and let those systems transmit back information directly to users' brains" (*Corrigan*). Again, we can see the bidirectional nature of such interfaces.

Elon Musk, the driving force behind Tesla cars, has formed a company called Neuralink with the goal of melding mind and machine through implanted technology. "Elon Musk lays out plans to meld brains and computers" (*Winkler*).

Mark Zuckerberg, founder of the social network Facebook, currently used by 3 billion people, has said that he plans to create a mind-machine interface within the next few years. "Facebook envisions using brain waves to type words" (*Frier*).

"Tesla's Elon Musk and Facebook's Mark Zuckerberg each aim to create the world's first brain-computer interface - devices that put the functionality of a laptop in your head" (*Brodwin*).

When this goal is reached, whether by DARPA, Musk, Facebook, or another corporation, humanity will be revolutionized. Just as typing on a keyboard produces letters on a screen or speaking to a smartphone causes it to execute commands, so specific actions of the brain will create electronic signals that can be used to control devices. When this technology is fully developed, instead of typing, we will just have to think.

Just think, and the thought will appear on the screen.
Just think, and the answers will download into your head.
Just think, and you can communicate with your friend.

Plugging in the human mind will have astounding implications. It will expand human capabilities to an almost unimaginable extent.

> We see the creation of technology that can meld the biological with the technological, and so be able to enhance human cognitive capability directly, potentially offering greatly improved mental [ability], as well as being able to utilize vast quantities of computing power to augment our own thought processes. (*Knapton*, quoting expert testimony before British Parliament)

GOAT TAGS

When brain-computer interfaces become widespread, computers will be melded to human minds, allowing people to have seamless interaction with the digital world. Cyborgs will have arrived.

Chapter 3
Dangers

A wise man fears and departs from evil, but a fool rages and is self-confident.

Proverbs 14:16

CYBORGS WILL have powerful advantages over ordinary humans - accessing vast information at the speed of thought, using powerful software to augment their brains, and harnessing powerful tools as if parts of their own bodies. They will be superhuman - even godlike - with amazing powers.

Yet cyborgs will not be gods. Their brains will be plugged into a powerful computer system which they will depend on in order to survive. That system will be their god, and cyborgs will be its slaves.

Cyborgs will face three primary dangers from this powerful system - the loss of privacy, the control of their minds, and the killing of their souls. Let's look at each of these dangers in turn.

1. Total Loss of Privacy

Throughout history, privacy has been a bulwark for humanity against oppression. Whenever people have been oppressed, they have been able to retreat to the inner sanctum of their own minds. The ability to resort to an unchained mind has been an enduring quality of humanity, preserving social progress when everything seems arrayed against it.

Mental privacy incubates revolutions. Paul spent years in Arabia, developing his radical theology under the inspiration of the Holy Spirit. When he began preaching, he shook the nation of Israel and ultimately undermined the Roman Empire. Martin Luther was sheltered at the castle of his protector, Frederick, as he thought and wrote, hidden from the armies of the Roman Church. Eventually, his ministry woke up Europe from the Dark Ages and sparked the Reformation.

Without mental privacy, these revolutions would have been impossible. If Paul's mind had been under a scanner, he would have been tracked down and arrested in Arabia by thought police years before he came to Judea and proclaimed the gospel of Jesus Christ. If Luther had been plugged in, he would have been discovered and flagged as a subversive long before he nailed the 95 Theses to the cathedral door. Without privacy, the Sons of Liberty, comprised of men like Samuel Adams, Paul Revere, and Patrick Henry, never would have been able to dump tea overboard in Boston Harbor and spark the American Revolution. Free, private minds are the death knell of tyranny.

Privacy is necessary for social progress. Without privacy, nations stagnate. The world needs truth, and truth often breaks into the world from minds that have been in touch with God in secret. Truth sets the oppressed free, which is why oppressors hate it. Oppressors try to eliminate privacy because they want to isolate their populations from truth in order to keep them enslaved.

Oppressive regimes support surveillance. In Nazi Germany, for example, mass surveillance was a fundamental policy of Adolf Hitler. The Nazis passed laws whereby "the right to privacy of communication by mail or telephone no longer existed" (*Schlingensiepen*).

According to experts, mass surveillance "is the single most indicative distinguishing trait of totalitarian regimes" (*Mass Surveillance*).

Dictators hate privacy. If a society has no privacy, only the current leaders can be the source of vast social change. Harsh rulers believe that loss of privacy will ensure their permanent dominance over their subjects.

We are already losing privacy at a rapid pace. Apps on our smartphones pile up personal data. Microphones sit in our living rooms listening to our conversations, transferring audio recordings to corporations. Whenever we go online, we give away our locations, emails, and photos to powerful organizations which suck up personal information like giant vacuum cleaners.

Online services often carry the illusion of being free, but we pay for them using the currency of our personal information. Whenever we use a search engine, we give away intimate details about ourselves, revealing our likes, sins, fears, and diseases. We pay as we go, and the more we use the internet— the more emails we send, the more we search the web, the more time we spend clicking away on social networks— the more our personal information piles up in the hands of powerful organizations.

The organizations which control our personal data are already extremely powerful. Corporations use our personal data to target advertising and maximize sales. Governments use it to monitor their populations. These organizations are able to connect the dots about our personalities and relationships in intricate ways, and they probably know things about us that we don't even know about ourselves. The power of these organizations increases in proportion to the amount of privacy we give up to them.

When brains plug into computers, there will be no privacy left. Thoughts will be constantly scanned, laid bare before an all-seeing technological eye.

The loss of privacy will have profound effects. It will mean the loss of freedom, the stifling of creativity, and rigid conformance. With no privacy left, cyborgs will be defined by the characteristics of conformity and dependence. In spite of their superhuman powers, cyborgs will actually be dull, gray cogs in a vast surveillance complex, powerless to change the system to which they are connected.

The complete loss of privacy will lead to perpetual tyranny and philosophical and societal stasis. Orwell's Big Brother will win, and the existing rulers will become permanent.

With the Antichrist on his way to rule over the world, the impending loss of privacy will be a human disaster.

2. Mind Control

The second grave danger that cyborgs will face is mind control. We have already seen how brain-computer interfaces are being developed to facilitate communication between humans and computers. Such interfaces will be bidirectional - not only allowing humans to access computers, but also allowing computers to influence the human mind. Uploads and downloads will speed information back and forth from cyborgs to computers along a two-way information highway.

> Scientists can do more with brainwaves than just listen in on the brain at work-they can selectively control brain function by transcranial magnetic stimulation (TMS). This technique uses powerful pulses of electromagnetic radiation beamed into a person's brain to jam or excite particular brain circuits (*Fields*).

As this technology develops, it will have dire results.

> New computers could delete thoughts without your knowledge, experts warn. New human rights laws are required to protect sensitive information in a person's mind from unauthorized collection, storage, use, or even deletion (*Johnston*).

This is a possibility now - it's just not widespread. "Scientists discover how to upload knowledge to your brain" (*Molloy*).

Such a radical discovery like this takes time to filter down into our lives, but it is only a matter of time before humans plug in, and through brain-computer interfaces, the content of their minds is altered. Just as the entries in online encyclopedias like Wikipedia are adjusted to reflect new information, so cyborg brains will be adjusted to reflect new information. And just as smartphones are regularly updated to improve their operations, so cyborgs will be updated so they can keep running smoothly. Data will beam into their brains, adding or deleting information, adjusting internal medical delivery systems, or altering the operation of bionic limbs. Updates like these will be a normal part of cyborg life. Mind control will be an essential and normal part of the cyborg experience, just as updates to our smartphones are normal today.

Burning Books

Throughout history, authorities have tried to mold the thoughts of their citizens. Books have been burned or banned since Gutenberg. Governments have attempted to purge their populations from "harmful" thoughts and get them to believe something "better". They want their citizens to embrace their political leaders, hate their nations' enemies, and reject "dangerous" religions or ideologies. This is not just stuff of history books. It's happened recently through force in Germany, Russia, and China, and it's happening through softer propaganda in almost all countries today.

The world is driven by propaganda. Through propaganda, politicians seek to secure dominance, corporations seek to sell their products, and ideologies seek adherents. Current efforts to purge the internet conversation from "wrong" information, "fake news," and hurtful content are new twists on the ancient goal of influencing people's thoughts.

It's one thing to remove information from a library shelf, school reading list, search index, or news feed. But directly removing "bad" information from the mind itself would take censorship to a whole new level. When brains plug in, "dangerous" ideologies can be rooted out, and "wrong" thoughts can be deleted. Political opposition can be carefully controlled, and reverence for the king or ruler will be ingrained. Faith in the chosen religion will be mandatory. Propaganda will stream directly into brains, turning people into model citizens who exemplify submission and patriotism.

The same connection that will give a cyborg access to vast information and superhuman abilities will become a leash, tethering it to accepted ideologies.

Criminalizing Christianity

Mind control, once it starts, will be very hard to stop, and it will be a particular threat to Christians.

Governments around the world are hostile to Christianity, and Christians are currently the most persecuted group of people on the planet. More Christians died for their faith in the 20th century than in all preceding centuries combined. The free practice of Christianity is blocked by many governments. Around the world, authorities ban Bibles and break up church services. Evangelism is often illegal, and converts to Christ are thrown into prison or put on death row. If these hostile governments could take it one step further, it's not hard to imagine that they would directly manipulate minds, making faith in Jesus impossible. Communist and Muslim governments are often ruthlessly antichristian, and today they rule over approximately 3.5 billion people. If these governments could, wouldn't they try to force the minds of their citizens to completely reject Jesus?

Turning people into cyborgs may be Satan's final attempt to block people from trusting in Christ and make hell a mandatory eternal destination for everyone. God preserves free will - in the Garden of Eden, he gave humans a choice of whether or not to stay in Paradise. Satan hates free will and wants to forcefully block people from choosing God. Turning people into cyborgs might be Lucifer's last great project upon planet Earth, his last-ditch attempt to destroy humanity forever.

3. Loss of the Soul

The third and most serious major threat that cyborgs will face is the loss of their souls.

Machines can move, think, and communicate, and they may be good at all of that. But no matter how intelligent an artificial system becomes, it will never truly be alive because it will never have a soul. This is the great barrier facing the advancement of artificial intelligence. Intelligent machines will always be soulless. Robots, no matter how convincing and apparently life-like they are, will always be *dead*. Without a soul, a robot is inhuman. It might be good at solving problems, playing chess, doing work, or even conversing, but without a soul, it will never be *alive*.

Humans are unique because they have souls. The human soul cannot be quantified or measured. It is a spiritual entity, beyond the realm of material science. The soul of a person can't be captured or replicated by a machine. A soul is much more than just a collection of information. Copying all the information within someone's mind and downloading it into a machine will not reproduce the person's soul.

Material science is good at many things, but it misses out on the most important things, which are spiritual. God is a spirit, angels are spirits, Satan and demons are spirits, and the human soul is spiritual. Spiritual things belong to a different realm - they can't be touched or measured - which is why science knows nothing about them. This is a distinct limitation for the fields of robotics, cybernetics, and artificial intelligence. Spiritual things are the most important because they will last forever. The human body will die, and the physical stuff it was made of will decompose, but the human soul will keep on going. After physical death, the soul will leave the body and go to another place while the body rots in the grave. The soul is eternal.

The soul *is* the person. Bodies don't die for lack of chemicals and molecules; they die because the soul has left them. Dead bodies (like that of Lazarus) come to life again not primarily because certain chemical reactions start happening again, but because the soul has reentered them. The soul defines the person.

Human souls are unique and invaluable to God because he can dwell in the human soul through the Holy Spirit.

In the quest to become stronger, smarter, and healthier, the machine part of cyborgs will tend to grow and grow, displacing the human part. This is because the machine part will be better at many things and last longer, satisfying the twin quest for greater abilities and longer life. Biological limbs will gradually be replaced by machine limbs. Skin will be updated with a better, smoother version. Organs will be replaced with something more efficient. Even the brain will be gradually replaced with a super-fast and efficient computer. Ultimately, the machine part of the cyborg might radically predominate, and the human part might disappear, leaving a soulless, mechanical body moving around on Earth. The exact point at which the soul would leave the cyborg might be difficult to tell, but it is too risky an experiment to try. The common image of cyborgs as unfeeling, inhuman machines is rooted in the fact that at some point, cyborgs will lose their souls.

Ray Kurzweil's insight bears repeating.

> We're going to become increasingly non-biological to the point where the non-biological [machine] part dominates, and the biological part is not important anymore. So even if that biological part went away, it wouldn't make any difference (*Kurzweil*, Fantastic Voyage).

Eventually, the human part of cyborgs will die, and the mechanical part will take over. Cyborgs are destined to become androids - dead robots.

Humans → cyborgs → robots.

When the human soul departs the cyborg, only an empty mechanical shell will remain. The cyborg will still move around, communicate, think, and maybe look the same. But it won't *be* the same. The human light in its eyes will have gone out. It will no longer be a person. It will be a mere machine. And without a soul, the cyborg will be cut off from God. A machine cannot be saved or born again. A machine will never have eternal life. A machine cannot be a habitation of the Holy Spirit. God dwells inside the souls of humans who are devoted to him. He doesn't dwell in temples made with hands, neither buildings nor machines. The ghost in the machine is not the Holy Spirit, it is the ancient dragon named Lucifer.

Cyborgs will become the mechanical corpses of people who have sold their souls for a few technological perks.

The implantation of technology will be an expensive exchange. It will give people the ability to interact seamlessly with powerful information systems, while simultaneously rendering them vulnerable to hacking and malevolent actions in ways merely biological humans could never be. Cyborgs will completely lose their privacy and become open to mind control by governments and powerful organizations. Gaining intelligence, strength, longevity, and amazing experiences through implanted technology will be at the cost of yielding to the control of computer algorithms, large informational systems, and powerful external intelligences. Plugging in the human mind will be at the ultimate cost of the human soul.

"What will it profit a man if he gains the whole world, and loses his own soul?" (Mark 8:36).

Chapter 4

Mark of the Beast

> And he causes all, both small and great, rich and poor, free and slave, to receive a mark on their right hand or on their foreheads, and that no one may buy or sell except one who has the mark.
>
> Revelation 13:16-17a

THE BIBLE SAYS that someday there will be one government over the entire world. Dominating every nation, it will be a platform for the Antichrist, a charismatic deceiver who will reject the true God and employ dark spiritual power. This government, like its leader, will totally oppose true Christianity. The groundwork for this government is already being laid by the United Nations and other supranational governing bodies.

This government is going to control humanity through the implantation of technology. The Bible says that it will put implants in people's right hands or in their heads. It is no coincidence that today most microchips for humans are currently inserted in the hand for ease of scanning, or in the head for ease of brain access. Implanted technology will be used for payment, communication, and control. Those who refuse this implanted technology will be unable to access basic financial, communication, or consumer services. No more internet. No more shopping at the local store. They will be unable to buy gas, electricity, or housing.

The Mark of the Beast is more than a way to buy or sell. It will be the connecting point between humans, the government, and the world economy. It will tag people for identification and control, plug them into a powerful informational system, and turn them into cyborgs.

Implementation of the Mark of the Beast won't happen overnight. Currently, we access the internet through external devices like laptops, and we pay for things using credit cards. The move toward internal technology will be a process. Like most new technologies, implants will gradually take over, as they make life easier, faster, and more convenient.

As people adopt internal technology, external devices will steadily become obsolete. Mobile phones and laptops will become artifacts, joining the cassette tape and floppy disk in the graveyard of antiquated technology. Eventually, it will be impossible to connect to the Internet using external devices. If you want to buy something, your credit card won't work anymore, and there won't be any cash. You'll have to get the Mark of the Beast.

The Mark of the Beast will turn people into Satan's cyborg slaves, blocking them from believing in God. This is why God prohibits his people from accepting this technology, and he warns that everyone who receives the Mark of the Beast will be consigned into hell.

> If anyone worships the beast and his image, and receives his mark on his forehead or on his hand, he himself shall also drink of the wine of the wrath of God, which is poured out in full strength into the cup of his indignation. He shall be tormented with fire and brimstone in the presence of the holy angels and in the presence of the Lamb. And the smoke of their torment ascends forever and ever; and they have no rest day or night, who worship the beast and his image, and whoever receives the mark of his name (Revelation 14:9b-11).

In the Bible, goats symbolize those people who reject God. At the end of the age, all humans who reject God are going to be tagged by internal technology. They will believe that these tags will help them. But these tags will actually mark them for eternal destruction and separate them from God forever.

Internal technology is Goat Tags.

Stop and Think

We, who are living on the brink of possibly the greatest technological change the world has ever seen, have a duty to ourselves and to future generations to step back, take a deep breath, and think about the implications. If we fling ourselves headlong into this new technological age of human machines without thinking, we may find ourselves unprepared to face the greatest challenge the world has ever known.

The course of the future may have already been set, and individuals may not be able to prevent the arrival of cyborgs. But it is not yet inevitable that you or I become cyborgs. As individuals, let us take responsibility for our future decisions, recognize the dangers of the current human trajectory, and prepare ourselves to refuse the implantation of technology.

In a few years, upgrading the human body with advanced technology will become the latest fad, and it may seem like everyone is doing it. When friends and family upgrade their minds and bodies and gain huge advantages over us, it will be hard to say no. Peer pressure to upgrade to the latest phone is strong, but pressure to receive implanted technology will be even greater.

Refusal

Refusing the Mark will detrimentally affect our careers, education options, and other important areas of life. But we must say "No," regardless of the consequences.

We need the mindset of Shadrach, Meshach, and Abednego, who were thrown into the burning fiery furnace but emerged unscathed. They did not fight. We are not called to fight physically, either. We are called to trust in God. The confession of these three bold Hebrews must be our confession:

> Our God whom we serve is able to deliver us from the burning fiery furnace, and he will deliver us from your hand, O king. But if not, let it be known to you, O king, that we do not serve your gods (Daniel 3:17b-18a).

We will not accept the internalization of technology. We will not become cyborgs. No matter the cost, even if we die, we will serve and worship Jesus Christ alone.

Refusing the Mark won't be easy. Cyborgs will race ahead with ever-increasing skills and strength, leaving merely biological humans behind. Ordinary people without implanted technology will slowly stop working in synchrony with the larger information system of the world. They will have no email, no internet, no banking, no credit cards, and no phones. They will be shut out of the informational system of the world, exiled from a system that no longer supports them. They will stand alone on Earth, surrounded by a swirling frenzy of digital activity that they are not able to understand.

Preparation

We need to prepare. God has not called us to be swept away by the tide of evil. He wants us to be like Noah, who prepared an ark, saving his family and, ultimately, the entire creation. It's time to think about the future and make radical decisions. Pioneering souls must be willing to risk everything to get to the place God is calling his people to be.

GOAT TAGS

Some keys to survival will be Christian community, growing food, obedience to God, and the supernatural power of the Holy Spirit. Spiritual technology from God is needed to defeat the evils of the Antichrist, and nothing else is going to work. (Please read the book, *Spiritual Technology*.) Small communities of saints, where believers are living together, working together, being holy together, and experiencing God together, will become the womb of Christ's manifested victory, which will ultimately sweep across the world.

Jesus was with Shadrach, Meshach, and Abednego to deliver them. He will be with his people in the end of the age to deliver them, too. God is the most powerful person in the universe, and he lives inside his people. He will have the final say, and his people will be more than conquerors in the end of the age. Christians need not fear the future, but can face it boldly, confident in the power of their Lord Jesus Christ. They can thrive, even without internal technology. They can win if they are filled with the Holy Spirit and submit completely to Christ.

Tower of Babel

There was a time in history, millennia ago, when technology was advancing at an alarming rate. All of humanity had gathered together into one place, united for a common purpose. They all spoke one language, had one common goal, and followed one leader. They were busy building a tower which would reach up to heaven.

The Tower of Babel was no ordinary tower. It was a symbol of man's rebellion against God, the enshrinement of a new religion to replace the old, and a mark of what humanity thought was their great progress apart from God.

God himself seemed concerned when he said, "This is what they begin to do, now nothing that they propose to do will be withheld from them" (Genesis 11:6b).

Whatever they think of, they will be able to do! Something was going on with this tower that meant humanity was crossing technological barriers that would keep nothing in their imaginations from becoming a reality. Their "progress" was speeding them ahead to an unknown destination, and they had no anchor in God or his word. They had reached dangerous waters which God could not let them cross.

Today too, we are confident in our technology. Futurist Arthur C. Clarke said, "Anything that is theoretically possible will be achieved in practice, no matter what the technical difficulties are, if it is desired greatly enough" (*Clarke*).

Technology seems to be only limited by the imaginations of its leading personalities. It almost seems that whatever Silicon Valley thinks of, they will be able to do. They are dreaming of colonizing space, upgrading their bodies and minds, and even living forever. "Ray Kurzweil wants to live forever. And like Peter Thiel and other Silicon Valley titans who have taken up the age-old search for immortality, he thinks he can" (*Blodget*).

It might appear that the growth of technology today is unique, and that we are crossing frontiers that have never been crossed before. It seems as if humanity is reaching into secret areas that have been off-limits until this privileged point in history.

But really, this is only a repeat of the Tower of Babel, just in a slightly different form. We have spiritually gone back to the plains of Shinar, where the great Nimrod was leading the human race in the wild construction of Babel's Tower, and where humanity was chasing dreams of usurping God's place and establishing utopia upon the earth.

And again, God is peering down from heaven to see what the children of men are building, just as he gazed downward at the massive tower that the entire earth was building with such frenzy millennia ago. He looks down, and when the advancement gets far enough, the rejection of Christ full enough, and the tower high enough, he will come down.

But this is where the parallels with Babel cease. This time it will be different.

Today's tower of technology won't just end with the division of languages. Differences in language can be overcome, and Google Translate is doing a good job of that.

Now it is the end of the age. Jesus Christ will come again, possibly when the Antichrist is at the apex of his power, possibly when a global tyranny threatens to annihilate all those who refuse to plug in and submit to it, possibly when divine hope seems to be lost. At the last minute, God will come and scatter all the dreams of carnal men and de-monized machines to the four winds.

This time it will be the final judgment. There will be no second chances. And the great dragon, that ancient serpent named Lucifer, whose writhing coils twist around men and nations, spurring them onwards and upwards in the feverish activity of building up spiritual Babylon, will be plucked from the sea of humanity and out of the earth, and thrown into the lake of fire forever.

Then the kingdom of God will reign. The glory of the Lord, washing over the earth, will cause the creation to resonate in a song of praise back to its creator. The full redemption of Jesus Christ will be revealed. Death will be no more, pain will be gone, sin will disappear, and life and health will spring up among all nations. The kingdom of light and life will prevail on the earth, and all remnants of the current human technological frenzy will be forgotten.

Jesus Christ will reign as King of kings and Lord of lords, and he will wipe away tears from all faces.

Therefore, since all these things will be dissolved, what manner of persons ought you to be in all holy conduct and godliness, looking for and hastening the coming of the day of God, because of which the heavens will be dissolved, being on fire, and the elements will melt with fervent heat? Nevertheless, we, according to his promise, look for new heavens and a new earth in which righteousness dwells (2 Peter 3:13).

Works Cited

Blodget, Henry. "Guess How Much Google Futurist Ray Kurzweil Spends on Food That Will Make Him Live Forever?" Business Insider, April 13, 2015.

Brodwin, Eric. "You can control this new software with your brain, and it should make Elon Musk and Mark Zuckerberg nervous." Business Insider, June 5, 2018.

Clarke, Arthur C. "Hazards of Prophecy: An Arresting Inquiry into the Limits of the Possible: Failures of Nerve and Failures of Imagination." 1962.

Corrigan, Jack. "The Pentagon Wants to Bring Mind-Controlled Tech to Troops." Nextgov, July 17, 2018.

David Axe. "The Pentagon's Wild Plan for Mind-Controlled Drones." Daily Beast, September 19, 2018.

Dearen, Jason. "Drones Fly Controlled by Nothing More than People's Thoughts." Associated Press, April 22, 2016.

Fields, R. Douglas. "Mind Control by Cell Phone." Scientific American, May 7, 2008.

Frier, Sarah. "Facebook Envisions Using Brain Waves to Type Words." Bloomberg, April 19, 2017.

Johnstone, Ian. "New Computers Could Delete Thoughts Without Your Knowledge, Experts Warn." The Independent [United Kingdom], April 26, 2017.

Kurzweil, Ray. "10 Questions for Ray Kurzweil." Time, Dec 6, 2010.

Kurzweil, Ray. *Fantastic Voyage: Live Long Enough to Live Forever.* N.p.: Plume, 2005.

Knapton, Sarah. "AI Implants Will Allow Us to Control our Homes With Our Thoughts Within 10 Years, Government Report Claims." The Telegraph [UK], Oct. 15, 2017.

"Mass Surveillance." Wikipedia: The Free Encyclopedia. Wikimedia Foundation, Inc. 22 July 2004. Web. 1 April 2019, https://en.wikipedia.org/wiki/Mass_surveillance.

Mims, Christopher. "A Hardware Update for the Human Brain." Wall Street Journal, June 5, 2017.

Molloy, Mark. "Scientists Discover How to Upload Knowledge to Your Brain." The Telegraph [UK], September 20, 2018.

"Research Priorities for Robust and Beneficial Artificial Intelligence." Future of Life Institute. Accessed Oct. 22, 2017. https://futureoflife.org/ai-open-letter.

Schlingensiepen, Ferdinand. "Dietrich Bonhoeffer 1906–1945: Martyr, Thinker, Man of Resistance." London, T&T Clarke, 2010, p.119.

Sheikh, Knvul. "Most Adults Spend More Time on Their Digital Devices Than They Think." Scientific American, March 1, 2017.

Winkler, Rolfe. "Elon Musk Lays Out Plans to Meld Brains and Computers." Wall Street Journal, April 20, 2017.

The Present Crisis

James Russell Lowell

Once to every man and nation
comes the moment to decide,
In the strife of Truth with Falsehood,
for the good or evil side;
Some great cause, God's new Messiah,
offering each the bloom or blight,
Parts the goats upon the left hand,
and the sheep upon the right,
And the choice goes by forever
'twixt that darkness and that light.

Then to side with Truth is noble
when we share her wretched crust,
Ere her cause brings fame and profit,
and 'tis prosperous to be just;
Then it is the brave man chooses,
while the coward stands aside,
Doubting in his abject spirit,
till his Lord is crucified,
And the multitude make virtue
of the faith they had denied.

By the light of burning heretics
Christ's bleeding feet I track,
Toiling up new Calvaries ever
with the cross that turns not back,
New occasions teach new duties;
Time makes ancient good uncouth;
They must upward still, and onward,
who would keep abreast of Truth.
Though the cause of Evil prosper,
yet 'tis Truth alone is strong,
And, albeit she wander outcast now,
I see around her throng
Troops of beautiful, tall angels,
to enshroud her from all wrong.
Careless seems the great avenger;
history's pages but record
One death-grapple in the darkness
'twixt old systems and the Word;
Truth forever on the scaffold,
Wrong forever on the throne,
Yet that scaffold sways the future,
and, behind the dim unknown,
Standeth God within the shadow,
keeping watch above his own.

www.bethelcornerstone.org